Crystals
and
Stones

The Joy of

Crystals
and
Stones

Connie Islin

Astrolog Publishing House

Cover Design: Na'ama Yaffe
Photographs: Ofer Ben-Mordechai, Liat Dor
Language Consultants: Marion Duman, Carole Koplow
Layout and Graphics: Daniel Akerman
Production Manager: Dan Gold

Astrolog Publishing House
P.O. Box 1123, Hod Hasharon 45111, Israel
Tel: 972-9-7412044
Fax: 972-9-7442714
E-Mail: info@astrolog.co.il
Astrolog Web Site: www.astrolog.co.il

ISBN 965-494-162-7

Published by Astrolog Publishing House 2002

10 9 8 7 6 5 4 3 2 1

Introduction

Crystals have played an important role in human development for millennia.

Since time immemorial, they have been used for communication, for healing, for ceremonial rites, for construction, as well as for surgery and increasing intuition. The shape of crystals is often reflected in the building forms adopted by ancient cultures, such as Assyrian ziggurats, Egyptian pyramids, and so on.

The legendary city of Atlantis is said to have been controlled by huge crystals, which served as a kind of communications network and power source. Royal leaders and experts in the occult have always used and consulted crystals in their daily actions. There are pictures of the construction of ancient South American cities, in which the foreman holds a stick with a crystal at the top of it; rays beam from the crystal toward the construction site.

Apparently the ancients recognized the power inherent in crystals, and harnessed it for their own purposes.

There is a theory that states that all knowledge and intuition, past and present, is stored in crystals, and it is a matter of tapping into these resources and retrieving the information from the natural data banks.

In a more modern context, each person has an individual crystal that can help him in daily life. One of the uses of crystals is as a pendulum, when a crystal attached to a thread is employed to solve problems, answer questions, and even locate mislaid objects.

The crystal radiates its own energies, and reacts with each person's energies - in this way, forming a kind of communicative team. Once this compatibility has been established, the person's life will improve beyond recognition, and he will go on to achieve things that he had not thought possible.

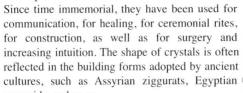

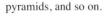

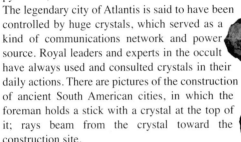

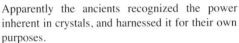

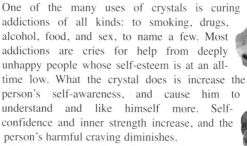

One of the many uses of crystals is curing addictions of all kinds: to smoking, drugs, alcohol, food, and sex, to name a few. Most addictions are cries for help from deeply unhappy people whose self-esteem is at an all-time low. What the crystal does is increase the person's self-awareness, and cause him to understand and like himself more. Self-confidence and inner strength increase, and the person's harmful craving diminishes.

Other uses of crystals are for enhancing success (as crystals have the capacity to promote considered, logical decision-making), and improving people's sex lives. Women can reach a physical awareness that enables them to achieve orgasm, and men can improve their virility. Crystals can protect us by absorbing the harmful radiation that bombards us from the environment. It is clear that the use of crystals is beneficial to human beings. Use this book to find your personal crystal - and watch your quality of life improve.

Crystals contain energy that influences every moment of your daily life, even if you are not conscious of it. While crystals and stones are often beautiful enough to be set in jewelry and expensive enough to be considered a good investment, they are endowed with other qualities and powers. With the help of this book, you can determine the appropriate one for your personal use. Crystals and stones can be worn as pendants, put under the pillow or in a pocket, or placed on their corresponding chakra to encourage the flow of energy. This comprehensive book presents the crystals and stones in alphabetical order, with a detailed description of the healing, mental, spiritual, energetic and predictive properties of each one. Let crystals become your best friends!

 Agate

 Alexandrite

 Amazonite

 Amber

 Amethyst

 Apache Tears

 Apatite

 Apophylite

 Aquamarine

Aragonite

Aventurine

Azurite

Azurite-Malachite

Beryl

Bloodstone

Calcite

Carbuncle

Carnelian

Cat's Eye

Celestite

Chalcedony

Charoite

Chrysocolla

Chrysoprase

Citrine

Coral

Diamond

Diopside

Dioptase

Dolomite

Dumortierite

Emerald

Epidote

Fluorite

Galena

Garnet

Gold

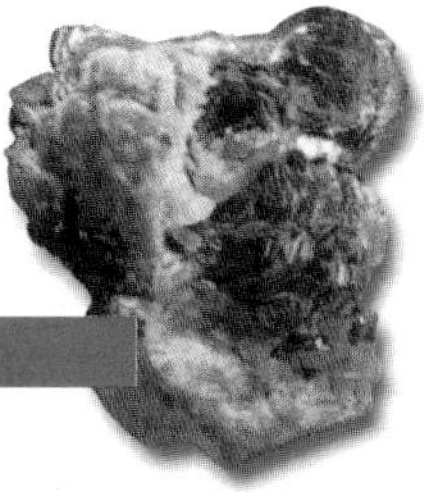

Hematite

Howlite

Hydrophane

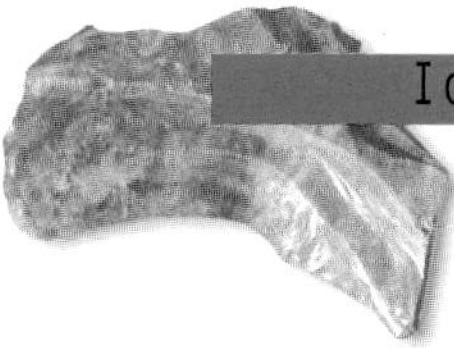

Iolite (Water Sapphire)

Ivory

Jade- Green

Jasper

Jet

Kunzite

Labradorite and Sunstone

Lapis Lazuli

Larimar

Lepidolite

Lodestone

Malachite

Moldavite

Moonstone

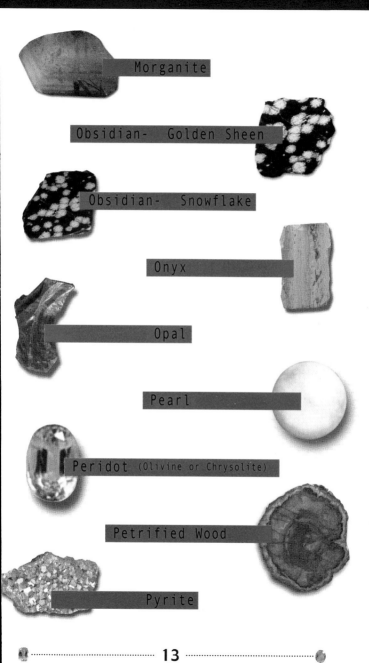

Morganite

Obsidian- Golden Sheen

Obsidian- Snowflake

Onyx

Opal

Pearl

Peridot (Olivine or Chrysolite)

Petrified Wood

Pyrite

Quartz (Crystal)

Rhodochrosite

Rhodonite

Rock Crystal

Ruby

Sapphire

Sard

Selenite

Serpentine

Smithsonite

Sodalite

Sugilite (Royal Lavulite)

Tektite

Tiger's-Eye

Topaz

Tourmaline

Turquoise

Agate

Agate exerts its strongest influence on intellectual people. Its effect is stronger on those who are guided by logic rather than by emotions or intuition. It is beneficial for those who conduct themselves in a reasonable and controlled manner and are not prone to mood swings.

Agate strengthens and directs the body and brain. It imparts a feeling of strength and power, courage and daring, and the ability to withstand pressure.

It can create in its bearer the determination to do things of which he would never have thought himself capable.

Agate's power is so great that it has strong and highly powerful healing properties.

Agate comes in a wide variety of colors:

Botswana agate: This stone is gray with white or brown stripes. It is soothing, and imparts courage, strength, and self-confidence. It provides protection against physical injury as it improves stability and the sense of balance.

Blue lace agate: This stone is blue with white or pale blue stripes. It neutralizes stubbornness and angry emotions as well as worries and feelings of frustration. It is associated with the area of speech, and should be held before giving a speech or even during regular conversation, so as to improve the ability to express oneself.

Turitella agate: This brown stone is considered to be a stone which attracts abundance. It is beneficial in situations where we must combine new and old, for example, new equipment in the workplace, a new partner or roommate, and the like.

Green agate: This is a stone which promotes happiness and sociability. It helps a person love others more easily.

Moss agate: This stone is green and contains what resemble plant fossils. The stone protects the heart in particular, both physically and emotionally. In the event of a fever, this stone cools the body down when placed on the forehead. The stone is particularly good for calming hyperactive children. It accelerates the growth of plants and seedlings.

Plume agate: This stone develops and encourages creativity. It helps people overcome shyness and reinforces sexuality.

Leopard skin agate: This stone imparts hope and fortifies the soul. It is good for

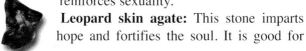

people who are in the depths of despair or for those who are in a state of depression or of mental tension, particularly one arising from angina pectoris. The stone balances the energies in the body, providing the strength and power to deal with fears of unclear origin. It is good for the development of awareness in people who are not sufficiently realistic. It is good for strengthening intuition.

"Black agate": This stone, which is formed from solidified tar, and is not genuine agate, has magical properties. It is considered to provide protection against the evil eye, and as such is used for making amulets or prayer beads. It is believed that when a woman burns this stone in the hearth of her home, the man she loves will come under her spell. The stone is therefore also used in love potions or as a love incense. This stone has the power to protect its bearer against dangers and to alleviate pain - particularly toothache, migraines, and epilepsy.

Alexandrite

Alexandrite amplifies happiness and joy and brings surprises. It stores up its user's energies so that they cannot be extracted from him. It also prevents its bearer from being affected by the negative energies of other

Crystals and Stones

people. This stone balances the nervous system and creates a connection between its holder's spiritual powers and his thought and emotion. By doing this, it promotes higher spiritual awareness.

This stone affords communication between the everyday, corporeal self and the Higher Self that guides the person in his wishes, desires, and abilities in life.

This stone is good for new beginnings; as such, it should be carried when moving to a new home, starting a new job, etc., as it symbolizes rebirth. It allows the bearer to take a deep breath, and provides energy for moving ahead.

Alexandrite promotes the balancing and opening of the energy channels between the base chakra and the heart and third-eye chakras. The stone also creates a balance between physical love and sexual attraction on the one hand, and spiritual love on the other. It gives its user the strength to transform his qualities and certain sides of his personality so as to enable him to fit in with society in the way he would like.

According to evidence from people working in healing, this stone can help cure leukemia and diseases associated with the circulatory system and the blood cells. This stone also helps to treat problems of the pancreas and spleen, and soothes the nervous system.

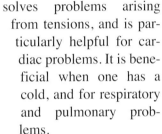

Amazonite

Amazonite is known for its properties of soothing and relieving emotional states. It balances the body, both mentally and emotionally, and is important for spiritual growth. It was even used in ancient Egypt.

This stone helps during childbirth, imparts a feeling of joy and happiness, and promotes proper and balanced development in children. It enhances the ability of infants to communicate, as well as their motor skills. This stone solves problems arising from tensions, and is particularly helpful for cardiac problems. It is beneficial when one has a cold, and for respiratory and pulmonary problems.

Amazonite symbolizes blooming, blossoming, and new beginnings, and it creates a balance between the individual and the universe. It helps achieve harmony between them and enhances the connection between man and nature.

A special property of this stone is its ability to balance a person's physical development with his emotional development, particularly during periods when there is a large gap between the two, such as during adolescence. It is very good for the development of children and babies. This stone can also help young people dur-

Crystals and Stones

ing the difficult period of adolescence, thanks to its ability to prevent self-destructive modes of behavior, such as addictions of various kinds.

Amazonite has characteristics that make it suitable for use in meditation: It provides a dimension of understanding and continuity between the world of the body and the world of the spirit. It has the power to dispel fears related to the experience of leaving one's body; as such, it is recommended for use with terminally ill people.

An additional trait of this stone is its ability to attract money and luck. This is why gamblers tend to wear it at every opportunity.

Amber

Amber was well-known as an amulet, and was used to protect the home and the individual. This stone releases tensions, eases stress, and soothes. It helps someone who is deliberating to make the correct decision and to

have the upper hand in situations of crisis. The stone is good for memory improvement and has a strong effect on the glands and the circulatory system.

Amber brings about harmony between people and nature, and with its help, negative energy can be released from the body, because it opens a sort of grounding channel through which those undesirable energies can be released. Because of its ability to neutralize negative influences in the body, it is considered to be a good healing stone. It is best worn as a wrist or ankle bracelet. It must be cleansed after each use, as it is a stone that absorbs negative radiation.

Crystals and Stones

Amethyst

This stone is helpful in balancing the energy centers in the body and in opening the third-eye chakra. It augments the feeling of courage, helps overcome fears,

Crystals and Stones

and enhances intuition and creativity. This stone helps people gain a deeper understanding of the soul, and develop greater self-awareness. It is good for meditation and helps its user make contact with the spiritual dimension.

This stone reinforces memory and neutralizes headaches, fatigue, and exhaustion. It releases energies of anger and resentment, and allows its user to show greater understanding toward those around him, as well as openness and forgiveness. The stone has great power when placed near the throat. It helps cure addictions to drugs, alcohol, smoking, sugar, etc.

The word "amethyst" in Greek means "non-drunk," that is, "against drunkenness." The Greeks used to wear amethyst when drinking alcohol, claiming that in this way, they did not get drunk in spite of imbibing vast quantities.

If the bearer or wearer of the stone is a male, the stone will make him more attractive to females. The amethyst is suitable for strengthening and ensuring the commitment between partners, and as such is an appropriate gift for a loved one.

This stone absorbs negative energies easily. It needs to be purified frequently; if this is not done, it can cause its bearer headaches and extreme irritability.

As its basic property is cooling, it is a good stone for use against fevers and inflammations. It can be used to cleanse the body of excess static electricity. It is an excellent stone for curing insomnia.

Crystals and Stones

Apache Tears

This stone imparts enormous energy to its user. It gives strength to the weak or to those who suffer from a lack of joie de vivre and cannot see life in an optimistic light. It balances those whose sexual energies are too strong - be they male or female.

It eliminates negative emotional blockages and enables

positive energies to flow, so that its holder is filled with feelings of enthusiasm, joy and sensuality which had been hidden deep inside him. It neutralizes feelings of deprivation or exploitation, and strengthens the ego and the spiritual and mental self. Apache tears are considered to bring luck and protection.

Apatite

Apatite enhances the ability to speak and express oneself; it eases stress, dispels tensions, and brings about effective communication between its bearer and other people. This stone strengthens the harmony between its bearer and nature and other people, and gives him patience and tolerance toward his surroundings. For these reasons, it is good for meditation.

This stone permits the flow of energy and information between the seven chakras, and reinforces the intellect, humaneness, and the willingness to offer help and to serve others. It is very good for people who are in the service professions, such as the various caring professions, providers of hospitality, and the like.

It is said that apatite helps strengthen the muscles, soothes rheumatic pains, and is effective in cases of osteoporosis. It fortifies the skeletal bones and the scalp, and is effective for all diseases associated with the bones. This stone removes physical and energy blockages, balances male and female energies, and enhances creativity.

It looks beautiful set in gold.

Crystals and Stones

Apophylite

Apophylite, especially those stones that are fan-shaped, opens the energy channels and chakras. When apophylite appears as a prismatic crystal, it operates by drawing light in from both of its sides. It develops the sensory system and intuitive understanding, and is particularly effective in strengthening the five senses.
Apophylite works mainly on the person's spiritual side. It helps its bearer feel more organized and capable of arranging his world in a structured manner, be this by

reorganization, by undergoing personal development in stages, or by constructing business or spiritual networks. This stone helps find ways to liberate the soul, draining off the negative energies within us in such a way as to help us understand what is oppressing us and what we no longer need.
Apopholyte stimulates the third eye and connects

between the spiritual, the material, and the physical. The stone helps create unity and wholeness, and prevents dichotomy and lack of wholeness. It stimulates the love for life and for all living creatures, plants, and nature in general. Thanks to this stone, the zest for life often returns and its significance becomes clear: It is worthwhile being alive. It stimulates the capacity for extrasensory understanding and safeguards against a drop in spiritual and mental energy and against energy being channeled off by another person.

Aquamarine

"Aquamarine" means "sea-water." From the physical point of view, it primarily affects the body's fluids and balances them. The stone is often used as an aid to dieting. This stone inspires love in people who engage in mysticism and are developing higher intuition, and is useful for meditation. The stone helps people who often feel ill at ease in company or in certain situations, and reduces embarrassment and confusion.

It enhances the bearer's speaking ability and improves his personal powers of expression and eloquence. It cleanses a person mentally and helps him purify his chakras. It affects the heart and throat chakras in particular.

It is good for banishing unjustified fears and helps students concentrate on their studies.

This stone alleviates toothache, problems of insomnia, blood vessels, stomach, and liver. It is beneficial for increased absorption of oxygen in the blood and for fighting off colds and pneumonia. It helps one find spiritual satisfaction in every act.

Because of the stone's soothing qualities, it can be used for calming the nerves and for good and harmonious flow of energies in the body. It provides protection to those traveling by air or by sea.

It deepens the communication and love between partners who wear it, and causes the person carrying it to receive a lot of attention from those around him.

Aragonite

This stone has a calming and balancing effect. It helps increase concentration and achieve mental peace, even in a noisy and busy place. This stone can soothe, especially in a situation in which a person must take steps in order to achieve efficient results, or if he must plan moves calmly and coolly.

The stone helps the bearer get through hard times more easily, without letting things plunge him into mental distress or depression. It prevents ulcers.

The stone helps the bearer reach solutions to complicated problems, find the knots and untangle them easily and effectively. It helps the person find the path necessary in order to implement the desired solution, to maintain existing frameworks, and to operate from within them. Regarding the physical qualities the stone radiates, it helps prevent hypothermia and induces a feeling of physical warmth.

Some people use the water in which aragonite was immersed (in daylight) to rub over the skin in order to prevent fungi and other skin ailments. Some drink the water to get over colds or to prevent them. In Africa the fluid is used to counteract hair loss and skin problems.

Crystals and Stones

Aventurine

Aventurine works on the heart area because of its characteristic gentle and stabilizing effect. It is good for prediction, increases the ability to function independently and to make significant decisions known to have importance in one's life, and helps ease mental pressures and physical tensions.

This stone helps unify mental and spiritual feelings and emotions so that with its help personal wholeness can be achieved. The stone is good for use by people of all ages. Physically, aventurine helps strengthen the muscular and nervous systems.

Crystals and Stones

Azurite

Azurite releases blockages in the brain and thereby arouses the consciousness and improves physical prowess.

The stone is also called the "Garden of Eden stone" because of its ability to soothe and bring about mental balance. The stone helps to unravel internal confusion and clarify thought patterns. It works well on emotional types who tend to follow the inclinations of their hearts rather than the dictates of reason.

The stone helps the bearer make decisions and eases feelings of destruction and of mental depression. This stone is used to relieve headaches, to get rid of migraines, to heal sore eyes, and to treat sinus problems.

It is known for strengthening bones, and is beneficial for healing skin and spleen problems. The stone is also efficient in activating the thyroid gland, relieving bone and joint diseases, and purifying the body of toxins.

This stone is best worn on the body as a pendant. Rubbing azurite increases the force of the energy it radiates. This stone aids in predicting the future, as it also enhances the ability to absorb knowledge from higher dimensions. It is also good for remembering dreams, and is customarily placed under the pillow in order to encourage predictive dreams.

Azurite strengthens spiritual forces and faith and consciousness. It is good to hold while meditating about past lives.

Azurite-Malachite

This is a combination of azurite and malachite, producing shades of blue and black. This combination creates a forceful and powerful stone, a stone with very strong properties. It enhances unique talents.

It works excellently with malachite, which is known for its properties of healing almost every bodily ill, as there is a positive interaction between the two.

The azurite strengthens the effect of the malachite, while the malachite enables the azurite to radiate feelings of harmony and consideration toward others.

Crystals and Stones

Beryl

Beryl is the name of a family of stones which includes a wide variety of gemstones: emerald, aquamarine, golden beryl, heliodor, bixbite, goshenite, morganite and a great many others.

Crystals and Stones

In general it can be said that beryl's properties relieve tension and help overcome feelings of tiredness, fatigue, or laziness. They increase mental balance and encourage spiritual cleansing. Beryl grants protection to its wearer and is good for personal use.

Bloodstone

Bloodstone helps with the healing and repair of the circulatory system. It is very dominant in everything connected with the circulatory system, particularly the organs associated with blood purification. It repairs the internal organs that purify the blood.

This stone cleanses the body of negative emotions and energies. The combination of red and green makes the stone very powerful. It is good for healing, and enhances self-confidence. It affords its user great strength, and its greatest effect is on the circulatory system. It increases curiosity, courage and mental

Crystals and Stones

balance. The stone's qualities lie in the fact that it calms volatile, "hot-blooded" people.

Because of its beneficial effect on the circulatory system, the stone is known for its importance during pregnancy, the menstrual cycle, or after an illness. It is also used in instances of injury, and women often wear it during menstruation. It is good for any problem having to do with circulation, such as iron deficiency, and problems with blood vessels, heart, kidneys and liver.

This stone helps its user to behave more assertively, to learn to stand up for his rights, and to demand his proper due. It neutralizes feelings of identification with situations of lack or shortage. It dispels fears regarding a decrease in assets or a loss of property.

According to legend, the stone was created from drops of Jesus' blood, or that of other tortured and martyred saints in other cultures.

There is another belief associated with bloodstone, that if the stone is placed in the middle of a sunflower, it transfers its power to the body and renders the person invisible, like a ray of light.

Crystals and Stones

Calcite

This stone works on the intellectual-mental plane. It obscures the boundaries of intellectual ability and thereby reveals aspects of the personality and abilities of which the person was not previously aware. Because of the stone's ability to soften the borders of the intellect, it is good for use by students who have difficulties studying or children who have trouble at school. It helps its bearer be more flexible in his opinions, and allows him to alter old thought patterns and ingrained opinions.

Calcite comes in several colors, each stone with its own characteristic qualities:

Red calcite: This stone has a significant effect on the heart and on the whole emotional system. The stone balances the energies that channel emotion, reinforces the person's emotional confidence, and prevents the accumulation of negative emotions in the body.

Green calcite: Neutralizes and releases fears buried in the subconscious, and eliminates pains. It balances the mind and imparts confidence to people or children who are facing a test. It stimulates the brain and prevents lack of concentration and momentary loss of memory (like in black-outs), causing the material being tested to be forgotten. The stone is good for healing allergies and purifying the body of toxins.

Orange calcite: This stone is good for strengthening sexual energies, particularly when it is placed on the second chakra. It inspires feelings of happiness and joy.

Blue calcite: This stone balances the person mentally and emotionally, and has a calming effect. It inspires serenity and tranquillity. It releases blockages and allows energy to flow freely and easily. It should be used in situations where the person feels the need to make changes in his life.

Pink/honey-colored calcite: This stone helps its bearer release old fears and anxieties. It inspires great optimism and is therefore good to use in moments of dejection and despair. The stone gives a feeling of confidence in anticipation of a new beginning, eliminates feelings of despair, and imparts hope. Physically, it is beneficial to the heart chakra.

Crystals and Stones

Carbuncle

This stone heals sores, helps prevent bleeding, removes bad influences on a person's mind, and dispels depression. This stone is known to preserve love over the passage of time and as such is recommended to be worn on the hands of both partners as a token of their long life together.

Crystals and Stones

Carnelian

This stone connects man to nature and helps him achieve harmony with the environment and ecological identification with nature. It is effective in treating fertility problems in both men and women.

This stone has a particularly strong effect on the emotions: It suppresses feelings of jealousy, neutralizes shyness and the desire to stay in the background, helps a person externalize his emotions and desires, and allows him to implement plans and ideas without any misgivings or fears of lack of success or failure. It imbues the wearer with courage and strength. This stone increases the ability to concentrate and augments the body's energies on all levels. It is excellent for people who are distracted or confused, or those with a poor memory.

It promotes creativity and imparts a good feeling toward oneself.

This stone makes it easier for a person to accept himself. It has an extremely positive influence on the entire lower abdominal region and strengthens the sexual energies. It is also beneficial in problems of digestion and eating, and for the prevention of fertility problems in women. It helps regulate menstrual periods.

Crystals and Stones

Cat's Eye

This stone is thought to bring luck. It is good for gamblers. According to Judaism, carrying this stone is a sign of wealth and a good life. Other religions believe that it is hard to keep secrets from someone carrying the stone.

This stone protects a person and fortifies his health. It eliminates negative energies from around its bearer and helps him be steadfast in his opinions. This stone is good for preserving the person's vitality and encourages him to maintain a fit and healthy body. It promotes the functioning of the internal organs such as the liver, spleen, and gall bladder.

Celestite

This stone has qualities and energies that connect it with the celestial world. It is particularly calming and radiates purity and serenity more than any other blue stone.

The stone allows its bearer to achieve a state of consciousness free from any tension and worry, a state in which thoughts are free of any ties and frameworks. The stone causes its bearer to reach a state in which awareness has a greater effect on his actions and gives him a feeling of confidence and control.

This stone can help people who suffer from a fear of heights or from claustrophobia.

Crystals and Stones

Chalcedony

This is a stone from which beneficial, positive things can be drawn. It reduces negative feelings or sensations of dejection or depression. It brings happiness and feelings of exaltation and joy to its user.

It helps the flow of positive energies and promotes openness and acceptance of the person's surroundings. It encourages a positive view of the world and everything around. It increases the desire for giving, and produces mental and spiritual balance.

It increases the ability to dream positive dreams that have a beneficial effect on the soul.

Crystals and Stones

Charoite

This is a New Age stone. It is good for dispelling fears that originate in the depths of the subconscious, fears that are connected with a person's spiritual side.

This stone will cause a transformation in conventional thought patterns and help the person shed them and penetrate to the true face of his personality without the defensive armor he has constructed against his surroundings. In this way, the stone releases feelings of guilt, fear, and sin that are engraved within him.

Crystals and Stones

Chrysocolla

Chrysocolla has wonderful positive energies that help balance the body and thought and generally balance various energies. It is good for emotional balance, so it is beneficial to hold it while experiencing emotional turmoil in order to calm and balance the mind.

Crystals and Stones

This stone alleviates and releases tension and distress. It is good for assuaging feelings of guilt when there is no solid or justified reason for them.

The stone is used for enlightenment and medita-

tion, and accompanies the time traveler on his journey through incarnations. This stone helps reduce the burden of heartache and hurt feelings.

It accelerates the process of new directions and beginnings.

It enhances the ability to see the positive side of any event and to overcome disappointments or loss.

The stone calms and imparts a feeling of space, freedom, serenity and hope. It is good for people with a tendency toward melancholy, or those who tend to be pessimistic.

Physically, chrysocolla rejuvenates and improves the functioning of the pancreas, the production of insulin, and the sugar level in the body.

Some use it for treatment of leukemia as well as for low white and/or red blood cell count. The stone helps relieve burns and fever and soothes asthma attacks. It helps respiratory function in general. (It is good for opera singers and people who need to use their full lung capacity.) It reduces discomfort resulting from ovulation and menstruation. It helps ease problems of metabolism.

The stone also helps with problems associated with the skeletal bones.

Chrysoprase

This stone's name means "fire of the house." The stone promotes happiness and inner calm. It helps to dispel egoistic feelings and suppresses jealousy. It is known to have an affect on the heart region and it works on balancing a person's emotions.

It strengthens inner vision and the ability to contemplate thought and body and to examine the self as objectively as possible. It is good for sexual balance and the healing of fertility problems in the reproductive system. It affects its bearer on both the spiritual and physical planes.

It is effective in the treatment of sleep problems and is considered a stone that brings good luck and wealth. It encourages success in business and important projects, along with success in matters of the heart, such as love and romance.

Crystals and Stones

Citrine

Citrine imparts energy and is good for self-confidence and a feeling of self-control. It encourages creativity and stimulates the ability for thought.

The stone imparts a feeling of stability and helps

Crystals and Stones

connect between logic and emotions which originate solely from intuition. It imparts the ability to communicate with others and the surroundings, releases inhibitions, and promotes rapid decision-making processes.

It is good for people who are hyperactive and who need their work-rate slowed down. It is recommended for children who suffer from anxieties or have difficulty falling asleep at night. It eliminates their anxieties and fears and prevents them from having bad dreams.

This stone enhances masculine energy and hence is good for increasing virility. Because of its color, it is connected with the sun and is very stimulating energetically. It awakens creativity and helps take advantage of every glimmer of creativity and originality in people.

This stone enables people to be more decisive and practical, and to listen to constructive criticism and accept it with love and understanding, without taking offense.

This stone gives great power to its user and imparts strong energies to him. It is better for people in the scientific professions than for those in the humanities.

Coral

Coral purifies and cleanses the crown chakra. It also helps balance the lower chakras in accordance with the power of the crown chakra. Coral balances the aura. It also helps to rejuvenate the cellular structure of the brain and to safeguard the bone marrow. It is beneficial in curing liver problems and jaundice. Its influence on the digestive system is very broad, and it can calm a person and get him out of a state of depression.

In the past, properties of protecting seafarers were attributed to this stone, and it was said to bring seamen safely back to their wives. The stone was also reputed to have properties of eliminating sterility; fertility was restored to women who wore jewelry made of coral.

Crystals and Stones

Diamond

The diamond symbolizes inner spiritual purity, strength, and wealth.

Many beliefs have sprung up about the diamond in all cultures because of its unique qualities. It does not diffract the ray of light that passes through it, nor does it reduce its quantity or strength; rather it transmits exactly what it absorbed.

The diamond is the first and most important healing stone of all the healing crystals.

This stone purifies the body and mind and helps eliminate energy blockages in the aura surrounding the human body. The stone releases tensions, angers, and sexual blockages; it is effective against the emotions of jealousy and envy, and it reinforces self-confidence and self-reliance.

The diamond operates primarily in the area of the higher chakras. It connects the third chakra with higher worlds and cosmic forces.

Because of the strength and power it contains, the diamond is capable of attracting numerous energies, and in cases of despair or negativity, the diamond is liable to attract negative influences. For this reason, it must be kept absolutely clean and pure.

It can be cleansed by soaking it for several hours in water containing a teaspoon of salt and a teaspoon of bicarbonate of soda.

Crystals and Stones

Diopside

This stone helps overcome the feeling of a lack of confidence. It is very good for healing and affects the heart chakra. It can help heal skin problems.

This stone strengthens powers of speech and expression. It can help when the bearer has to give a speech in front of people or when he has to express something important in writing.

Crystals and Stones

Dioptase

This stone is particularly good for healing heartfelt worry or traumas created by emotional betrayal. It works on the heart chakra, healing and cleansing it. It helps heal emotional scars from the past, especially ones resulting from abandonment, be it romantic abandonment or another form of abandonment, such as abandonment by parents due to divorce, death, etc.

This stone helps its user regain the feeling of basic security that he lost, and reinforces energies of love and the ability to trust and love.

If the stone is placed on the third-eye chakra (just above the spot between the eyebrows), it will help dissipate guilt feelings.

Crystals and Stones

Dolomite

Dolomite leads its bearer to inner serenity. It influences the lower chakras, meaning the stomach region, intestines, reproductive area, kidneys and liver. It releases tensions, and affords mental and spiritual rest. It is beneficial in situations in which the person faces a test or a certain stage in life that is to determine his future, causing him tension, stomach pains, and muscle spasms. In situations like these, the stone will lead to calmness and a feeling of confidence and control of the situation.

Dumortierite

This stone is good for someone who needs all the information he has acquired, be it in this life or in past lives. It allows all the information accumulated to come to the fore so that it can be used in everyday life when needed.

This stone works on the memory center in the brain and removes the barriers to the information stored there. The stone encourages the application of a practical and businesslike approach to life. It encourages a utilitarian approach while listening to the self and cultivating the ego.

Crystals and Stones

Emerald

This stone possesses properties pertaining to the mental, physical and chemical planes. It is soothing, and it improves the power of expression, speaking ability, and eloquence. It cures diseases of the intestines, back and heart. It helps reduce tension and improves vision when one focuses on it (a trait common to many green stones).

The emerald neutralizes and balances feelings such as lack of self-esteem or alternatively, exaggerated self-esteem, both of which are contrasting characteristics of the same energy. It is a good stone to wear at times when a person has to make decisions about his future path. It is also called the "truth stone," and the bearer can tell by intuition during a conversation with someone else whether the other person is telling the truth or lying. This stone helps those who find it hard to give or receive love to break the barriers and reservations surrounding the heart, and to accept things more easily and joyfully.

This stone draws its power from the sun, so it should be "warmed up" occasionally in the sun. The emerald has always symbolized love, abundance, generosity, good-heartedness, triumph over sin, and supreme morality. It helps its bearer achieve mental, emotional, physical and spiritual balance. This is why it is considered a most important stone for healers.

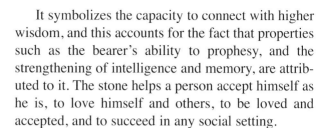

It symbolizes the capacity to connect with higher wisdom, and this accounts for the fact that properties such as the bearer's ability to prophesy, and the strengthening of intelligence and memory, are attributed to it. The stone helps a person accept himself as he is, to love himself and others, to be loved and accepted, and to succeed in any social setting.

Epidote

Epidote is known as a stone with properties that inspire self-confidence in its bearer. It balances and nourishes sexuality. It enhances the feeling of self-love and self-esteem. It helps in various cures; in order to make the most of it, one should seek the aid of a healer, and not use it for healing purposes on one's own.

Crystals and Stones

Fluorite

This is a very good stone for meditation. It cleanses the highest chakra and directs the body's energies. It balances between body and mind. It induces enhanced awareness.

It helps a person reveal himself to the mental and spiritual layers within himself.

It is effective for states of spiritual turmoil. It is a good stone for channeling and for curbing excess energies or undesirable energies appearing in the aura.

This stone enhances the ability to concentrate, as well as the possibility of realizing childhood dreams by strengthening the will to do so.

One of its components is fluoride; thus it is known to play an important role in strengthening the teeth and the bones in the body. It is also effective against rheumatism, lower back pain, injuries, and inflammations of the joints.

Crystals and Stones

Galena

Galena reinforces self-confidence and recognition of self-worth. The stone dispels feelings of missing out and of emptiness. It protects the bearer against negative moods and pessimism. It encourages him to return to his true self and his inner truth. It purifies improper intentions and eliminates bad thoughts. The stone has a physical effect on vision, and a calming effect on the nervous system. Galena strengthens the imagination and helps to liberate hidden thoughts which have never come to the fore.

Crystals and Stones

Garnet

Garnet is actually a family of stones that affect a person mainly in the mental and spiritual areas. This stone is good for people lacking confidence. It soothes nightmares. It influences the world of the imagination and dreams, enabling the dreamer to remember them, and encourages awareness in dream interpretation.

This stone brings enlightenment, and purifies and strengthens physical energies. It is beneficial for headaches and backaches, and helps relieve neck pains.

In general, it can be said of this stone that it strengthens the body and affords physical strength and immunity. It is used to correct problems of low blood pressure.

The stone promotes motivation, courage, and energy. It improves the self-image and the social image.

Garnet is considered to attract success and abundance. It protects its bearer from negative events and neutralizes fears and weaknesses.

Gold

Stones that contain the mineral gold increase the energies of the physical body. They cleanse it and enhance its absorption of the sun's energies, meaning the masculine yang force. Gold balances the heart chakra and strengthens the nervous system. Having too many stones containing gold near a person can provoke outbursts of violence toward his surroundings. Be careful!

Crystals and Stones

Hematite

This stone has magnetic properties. It has a great influence on people undergoing treatment. The stone absorbs enormous energy from the universe and imparts it to its bearer. It creates a balance between body and mind. It is known to have a marked influence on the circulatory system. It stimulates blood

Crystals and Stones

circulation and purifies the blood. It is good for people with anemia, and from current reports, it is found to have a positive effect on AIDS patients. (It should not be carried during menstruation because it stimulates the circulatory system and is liable to cause excessive bleeding.)

The stone provides protection against negative emotions, and increases resistance to pressure. It is good for rational, logical people who do not rely on their senses and intuition, as it increases intuitiveness and the ability to sense what is taking place beyond the visible and obvious.

The stone reinforces and increases fighting spirit, determination, will power, assertiveness, courage, the ability to cope with crises, and the ability to be optimistic and to transmit this to one's surroundings. It increases the person's ability to be steadfast in his opinions and to demand his due when he knows justice is on his side. Because it helps a person justify himself to others, it was considered the stone of the "throne of judgment" during the time of the Roman Empire.

Crystals and Stones

Howlite

This stone has the effect of inspiring one with harmony and blending energies. The stone has a calming and balancing influence. It is beneficial in situations of stress, worry, and anxiety. The stone helps thought and brain function. It is calming when thoughts are on the rampage in a person's head, giving him no rest.

It regulates the forces stored in thought energies, and channels them toward new ideas and thought clarification. The stone helps to reveal creative ideas, and is good for enhancing the creative imagination. With its help, original thoughts that had previously not come to the fore or been actualized can be implemented.

❀ Crystals and Stones ❀

Hydrophane

This stone has a transparent white color. Its major power lies in preserving love. The stone's qualities are mainly courage (the stone dispels fears and boosts self-confidence) and seeing the bearer's future.

Iolite (Water Sapphire)

This stone enhances self-confidence, instilling in the bearer the feeling that he can "take command."

The stone gives its bearer a feeling that he is no less worthy than others. It is good for people who value the opinions of others much more than they do their own, and who suffer to some degree from feelings of inferiority. The stone will instill a feeling of confidence in such people. They will come to recognize their self-worth and will learn to rate themselves more objectively vis-á-vis their surroundings. The stone grants its bearer a broader point of view, so that his world is not limited to a narrow area; rather, he is able to enjoy a world view that stems from perspective and overall vision.

Ivory

Ivory's influence is mainly in the spiritual area. It allows one to see life more lightheartedly and helps develop personal awareness. With it, one can achieve spiritual enlightenment.

Crystals and Stones

Jade - Green

Jade imparts illumination to its bearer, as well as serenity and harmony between body, mind, and soul. In South America, its Spanish name is "Piedra del Lado," which means "the side stone." This name stemmed from the belief that it could cure irregularities in the kidneys if placed on the sides of the body.

The stone has always been popular in the Far East, where it was used to strengthen the body; it was also believed that its properties could increase male

Crystals and Stones

virility and fertility. In China it is believed that jade contributes to the five main qualities necessary for a person: courage, justice, compassion, humility, and wisdom. There it is considered an "obligatory" stone. When struck, it emits a unique tone which, according to the Chinese, was the sound of lovers.

When the stone is held, it emits a feeling of calm and serenity.

It helps in repelling negativity and in getting through periods of crisis smoothly and easily.

This stone is excellent for those who are unable to express themselves emotionally. It inspires divine and selfless love, and promotes the development of an altruistic nature.

It is beneficial for eye, heart, throat, and kidney problems. Jade helps its bearer reach correct decisions. The stone is thought to ensure longevity and helps in dealing with "no choice" situations or situations in which the person reaches a dead end. Drinking water in which jade has been immersed induces calm, strengthens the bones and muscles, and purifies the blood.

Crystals and Stones

Jasper

This stone has a cumulative quality. Although its effect is gradual, it is long-lasting. It strengthens endurance and the ability to withstand difficulties for extended periods. It is a stone that symbolizes patience, tolerance, and perseverance.

Crystals and Stones

It is good for developing physical fitness and tolerance. Jasper is calming and has qualities that bring its bearer mental relief. It helps release tensions and balance body and mind, conscious and subconscious. Jasper is considered to be a good healing stone. Its effect is gradual and stable. It works on the lower chakras- the digestive system, the pelvic area, the

intestines, and kidneys. It also helps cure skin problems and reinforce the muscles.

There are a number of different shades of jasper, ranging from brown to red, which have their own characteristic qualities:

Red jasper: Imparts strength and power, as well as the ability to cope with difficult situations, or situations that involve withstanding physical hardships. (It is good for soldiers in boot camp, for example.) This stone releases blockages in the liver. It helps the bearer withstand negative influences that could affect the emotions. It protects against accidents and the evil eye.

Yellow-brown jasper: Helps its user to be more realistic and grounded, as well as to avoid emotional involvement with people who have a negative influence. This stone balances the hormones in the body.

Picture jasper (brown): This stone increases tolerance and eases tensions. It causes its user to contemplate past mistakes and avoid repeating them.

Jet

This is black organic matter, resembling coal, which primarily alleviates feelings of distress, dejection, and depression. It is a good calming agent during periods of stress, and it relieves emotional overload. It strengthens the spirit in times of trouble and brings about the release of mental and emotional blockages. This stone enhances survival ability and grants its bearer mental strength.

Crystals and Stones

Kunzite

This stone is good for use by people in the process of withdrawal from addictions such as smoking, drugs, alcohol, food, etc. It helps overcome fears, anxieties, and feelings of despondency and depression.

It enhances self-awareness and intensifies mental/emotional ability. It enables a person to accept situations of giving selflessly and of submission more easily.

It works strongly on the heart chakra and helps the bearer open up mentally and emotionally with those around. It increases unconditional love and heightens sensitivity toward others.

Labradorite and Sunstone

These stones bring optimism and joy to their bearer. They eliminate negative feelings, intensify the positive energies in the body and even recharge it with new energies. They are good for people who work long hours and require strength and energy.

Crystals and Stones

Lapis Lazuli

This very powerful stone was the most common gemstone in Egypt during the time of the Pharaohs, when its value was equal to that of gold.

It has an excellent effect on the ability to think clearly. It is good for people who find themselves at a crossroads in their lives and must decide on their future.

It helps people reach decisions in a neutral and objective manner.

It stimulates inner powers that the person was not aware of possessing. Because of this, it is a very good stone for meditation. When carried, it stimu-

Crystals and Stones

lates the brain to produce creative energy. It is good for anyone who requires a push and needs to have his creativity strengthened.

This stone dissolves negative energies, helps to dispel gloom, and eliminates anger and irritability. It improves both written and oral powers of expression. It awakens such strong energies that its bearer should know how to use it wisely and treat it cautiously.

This stone affects many systems in the body, such as the bones and the respiratory system (primarily throat and lungs), and it is excellent for cleansing the circulatory and immune systems.

Crystals and Stones

Larimar

This is one of the New Age stones. It connects feelings and emotions with thoughts. It is used primarily by people who engage in channeling. It is good for those who are trying to be part of nature and those who are capable of communicating with animals and the environment (for example, dolphins, plants, etc.).

Lepidolite

This stone contains a combination of pink (heart) and purple (head). It is important to use this stone according to its colors.

When it tends to pink, the stone soothes and eases heart problems.

When it tends to shades of purple, the stone has an excellent effect on over-active people who need to be calmed and slowed down.

This stone balances emotions, brings about mental stability, and neutralizes anger and anxieties. It reinforces spiritual capability and affords communication with higher worlds. It is beneficial in the treatment of all muscle pains and for improving blood circulation.

Crystals and Stones

Lodestone

This stone is used primarily to enhance speaking, speechmaking, and the ability to express oneself freely. The ability to respond improves with its help. The stone has a calming nature and can help people whose sleep is disturbed or restless.

Crystals and Stones

Malachite

This is the "sister" stone of chrysocolla. It provides protection against environmental contamination and is therefore recommended for people who live in areas tainted by radioactivity, or near areas which are dangerous as a result of toxic gases. This

stone balances and strengthens the brain, and alleviates sensations of tension, irritability and confusion. It heals blockages of energies that affect emotion and releases the person from emotional traumas.

Malachite is a highly protective stone, which intensifies energies, be they positive or negative, so it is important to know when to carry it. If for some reason you do not feel comfortable with it, it is better not to carry it, as it will intensify any negative emotions.

Crystals and Stones

This stone has acquired a reputation of being one that attracts good luck and abundance. Because of its property of absorbing negative energies, it is beneficial in healing any part of the body. It can be placed on any part that hurts, and there will be an immediate feeling of relief. Because of this, the stone should be purified frequently.

Malachite is good for curing allergies, lower back pain, mental diseases, and diseases of the spleen, pancreas, and thyroid gland.

(Malachite-Lazulite)

This stone imbues its user with mental serenity and calm. It intensifies the feeling of being able to cope with challenges, and imparts courage. It strengthens and nurtures the male ego, but also inspires harmony and contentment, according to how it is used.

Moldavite

This is an excellent stone for initiating communication with higher worlds. It induces enlightenment and is good for meditation. It works very strongly on the sixth sense and on broadening awareness, and reinforces telepathic abilities.

Moldavite possesses extremely high energies which permit it to open blockages in every one of the chakras. It balances the body physically, and enhances its bearer's intellectual and mental abilities.

Crystals and Stones

Moonstone

This stone has the ability to make its bearer exercise greater control over his emotions. It enables him to respond to events with balanced judgment rather than with uncontrolled emotional outbursts. It balances his male-female energies and hence helps solve many problems relating to love.

It is very good for creating harmony between partners, especially those who have been married for a long time. It stimulates the positive energies between them.

This stone enhances intuition and heightens sensitivity toward others. It is calming, and it ensures that there are no sleep problems.

The stone prevents a person from experiencing extremes of moods. For this reason it is good for teenagers to carry. It balances the extreme reactions to which this age group is prone, and instills in them a feeling of confidence and control of their lives.

Moonstone soothes pain and primarily affects the respiratory system and lymph glands. It heals ailments of the spleen, pancreas, and pituitary gland.

Crystals and Stones

Morganite

This stone influences the area of love. It intensifies its bearer's ability to love both himself and others, and as well as his surroundings. This stone is good for people who devote themselves to civic work, such as volunteers in various organizations, or philanthropists.

This stone cleanses and energizes the heart chakra and the third eye. It has a beneficial affect on the female reproductive system.

Crystals and Stones

Obsidian - Golden Sheen

This stone is a natural glass. It therefore constitutes a sort of reflection of one's existential situation, and it is influential in altering emotions. It is preferable not to hold it when feeling upset or unstable.

The stone's golden color enhances whatever the bearer is feeling and intensifies emotions. Holding it when experiencing emotional or behavioral vacillations should be avoided.

Crystals and Stones 🌿

Obsidian - Snowflake

This stone is a natural glass. It is black in color with white "snowflakes" in it, and radiates great power. It imparts a great deal of self-confidence and strength and energy to its bearer, and is capable of transforming the internal feelings of its bearer from

negative to positive. It also has the power to reduce over-sensitivity and to balance energies.

It is good for times of physical or emotional changes, such as moving to a new area or a new home, or a change in emotional status as a result of deprivation.

It is beneficial in releasing emotional blockages, and allows a good and easy flow of personal strength.

On the physical plane, this stone strengthens and balances the intestines. It is good for when the stomach is sensitive and for stomach-aches caused by over-excitement. It strengthens the stomach and the muscles, and endows them with strength and vitality.

Onyx

Onyx, one of the stones in the biblical High Priest's breastplate, is considered a "smart" stone. It relieves tension, balances male-female polarities, enhances hormonal balance, increases the courage of its bearer, and increases his self-control.

It also helps its bearer form sincere and loyal friendships.

This stone creates discernment and objectivity in its bearer. It contains a high degree of inspiration, and it is said to be a stone that "remembers," meaning it absorbs the experiences and impressions of its bearer, and, through extrasensory ability, the experiences it absorbed can be retrieved from it.

There are several types of onyx, each of which has its own unique additional properties:

Pale onyx: By wearing it or carrying it on the body, a spiritual path is initiated without its wearer even being aware of it.

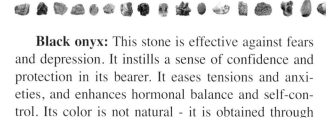

Black onyx: This stone is effective against fears and depression. It instills a sense of confidence and protection in its bearer. It eases tensions and anxieties, and enhances hormonal balance and self-control. Its color is not natural - it is obtained through chemical means.

Blue onyx: This color is also created chemically. It helps its wearer or bearer become more assertive, and to say what needs to be said in a more resolute and less emotional manner.

Green onyx: This stone, also chemically colored, works on the emotions and constitutes an emotional healer.

Onyx-agate

Onyx-agate has the property of attracting good luck, so it is good for gamblers. The properties of protection against the evil eye, spells, and evil spirits are also attributed to it. If you work in a job together with a partner, it is recommended that you both wear the stone. This will prevent you from having needless disputes and quarrels, and will preserve the harmony between you.

Crystals and Stones

Opal

The opal intensifies sensations. If you feel ill, the opal will be affected and its color will almost disappear. This stone leads its bearer to a high awareness of his emotions and enhances intuition. If worn at the center of the neck, the stone will bring about the opening of the fifth chakra.

It enhances creativity, love, happiness and joy, and is good for creating sexual relationships. It intensifies sexual attraction and increases the interest in sex. The stone helps its bearer use logic, particularly when solving differences and problems which require cold, rational analysis. This stone also strengthens physical energies and helps the bearer or holder to concentrate and be alert.

The opal contains water, and is therefore considered a stone that stores up radiation and energy powerfully. As a result, it radiates very strong energy that can cause distraction, confusion, and the inability to concentrate.

It is best worn on the pinkie, far from the body. It should not be carried or worn together with other stones. Because of its diffusive property, its non-

Crystals and Stones

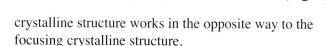

crystalline structure works in the opposite way to the focusing crystalline structure.

Some say that if the stone is given as a gift in an act of good will, it brings good luck and cultivates feelings of friendship.

Because of its receptive nature, the stone was used in the making of amulets and as a "wishing stone," particularly in matters of love.

Legends and beliefs that the stone also cures eye diseases sprang up around the white opal, because of its bright white color and its similarity to the white of the eye. Fire opal, on the other hand, is considered to be associated with the blood because of its color.

Crystals and Stones

Pearl

Although there are many different kinds of pearls throughout the world, their energies are the same. The pearl symbolizes purity, and is therefore considered to be a stone that purifies the person's soul.

The pearl effects harmony and balance because of the fact that it contains negative ions. Excessive positive ions in a person's proximity cause him to feel that his emotional balance has been upset. The pearl balances the emotions.

It is good to use in cases of over-excitement or intense emotions. It instills peace and helps restore serenity and control when the person is distraught.

It is good for the treatment of breathing difficulties in cases of asthma, and for the treatment of the lungs and diaphragm.

Crystals and Stones

Peridot (Olivine or Chrysolite)

Peridot enhances intuitive sensations and the ability to control oneself. This stone helps one reach a high level of contact with higher worlds and brings man closer to his divinity. It is very good for use during meditation, particularly when there is a feeling of straying from one's spiritual path. It is a pure stone that is of great help in eliminating negative emotions and bad thoughts. It assists in intensifying the spiritual forces from within the inner self, and helps increase personal confidence. It helps in foretelling the future and in divining fate.

Peridot is good to hold when feeling sad. It imparts optimism and clear, lucid thoughts. The stone is associated with good luck and positive adventures, so it should be carried during periods of transition.

It is effective in curing problems of the spleen, pancreas, and gall bladder. Drinking water in which peridot has been immersed purifies the liver.

Petrified Wood

This stone is thought to be the bearer of good luck. It is connected to the forces of nature, and makes it easier for people who do not have both feet firmly on the ground to be more grounded and less "spaced-out." This stone dispels fears and helps the bearer cope with daily reality. The stone imparts energy and vigor to its user. It is good for exhaustion, fatigue, and heaviness. The Indians would use it against infections, injuries, and accidents. It is considered to be a stone that strengthens and is beneficial in treating joint inflammations, problems in the bones, and a variety of infections.

Crystals and Stones

Pyrite

This stone uplifts the spirit, and helps one get into a good mood and relate to unpleasant incidents lightly. It is good in instances of low spirits, dejection, or depression. It helps create trust in others and harmony with the environment. It has a beneficial effect on oxygen absorption in the body and on blood circulation. It helps alleviate problems in the digestive system and enhances brain activity.

❧ Crystals and Stones ❧

Quartz (Crystal)

There are a great many quartz stones. The unique formation of quartz crystals is what gives the stone its power. There are quartz stones which are classified according to shape: single point, double point, barnacle quartz, and more. And there are stones which are classified according to color: milky, smoky, rutilated, rose, etc.

The list of common quartz stones contains around sixty types. The most common of them are: smoky quartz, rose quartz, and snowy quartz.

Quartz serves to intensify electrical impulses. It has the property of absorbing and conducting radio and electromagnetic waves. This is why it is used in the manufacture of watches and computers.

Clear white quartz: This is the most basic and important stone of all the gemstones, both for healing and for spiritual development. Clear transparent or pure white quartz promotes the penetration of positive energies into one's thoughts and emotions. The stone permits objective viewing of situations in life and helps a person get out of negative habits.

This is a very good stone for healing, as it imparts energy and transmits healing energies exactly as they are, without distortion or alteration. It balances the chakras and heals various ailments of the body.

It is particularly good for healing and strengthening the intestines, stomach, and pineal gland. This stone intensifies every feeling. For this reason, it should not be carried when the bearer is feeling low, depressed, or despondent.

It enhances higher awareness and clarity of thought. This stone allows one to reach deep meditative states, and through it one can make contact with one's Spiritual Guide or Higher Self, and one's inner self.

This stone energizes and instills a feeling of vitality and the desire to experience as much as possible of what the universe has to offer.

Rutilated quartz: Symbolizes renewal and rejuvenation. It sharpens the ability for thought, imparts a feeling of vitality, and fends off the effects of aging.

It balances the body's energies and enhances the power of the third eye. Physically, this stone helps the digestive system, eases intestinal problems, and promotes the absorption of food and easier digestion.

Blue quartz: Enhances originality and creativity. It intensifies the person's mental and spiritual abilities and connects between these and the third eye.

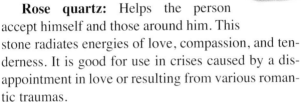

It enhances the feeling of calm and serenity. It is best worn high up on the body, as a pendant or earring.

Rose quartz: Helps the person accept himself and those around him. This stone radiates energies of love, compassion, and tenderness. It is good for use in crises caused by a disappointment in love or resulting from various romantic traumas.

It restores the person's ability to love without being filled with bitterness, and makes him feel whole, calm, and tender. This stone is very effective in the healing process because of the energies that flow from it and because of the fact that it does not counteract the action of any other stone.

It strengthens the cardiac system and is beneficial to the heart chakra. It is also good for healing of the digestive and circulatory systems and the reproductive organs.

It can be worn without restriction, because of its transparency, which is useful in any situation and for every purpose and objective.

Red phantom quartz: Very few stones like this have ever been found on the face of the earth. It is known for promoting awareness and spirituality as well as the combination of the spiritual and the earthly, of the upper chakras and the lower ones.

Smoky quartz: This stone has a unique influence on a person's spiritual side. It helps eliminate nega-

Crystals and Stones

tive energies or undesirable influences. If you wish to dispel negative thoughts or get out of unpleasant situations, you can be helped by it.

This stone enhances inspiration and the ability to foretell the future.

It is effective during astral travel and helps a person connect to the energies of the earth, nature, and the environment.

This stone helps with problems of the reproductive system, the lower back, the heart, and the nervous system. It is useful for cancer patients who are undergoing radiation or chemotherapy, because of the radioactive radium it contains.

Snowy quartz: As opposed to the regular quartz stones, this stone does not conduct electricity. However, it represents purity at its most sublime level.

It helps to link up to a higher force and to unite with God and with higher worlds. It develops extrasensory ability, influencing the third eye and rousing it to action. This stone intensifies the feeling of divine love, peace, and wisdom.

Rhodochrosite

This stone is good for strengthening self-aware-ness and merging it with the subconscious. It helps its bearer overcome fears stemming from feelings of dejection or paranoia of all kinds. It is very calming, induces sleep in states of tension, and helps a person see things in a rosier light.

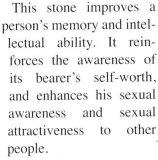

This stone improves a person's memory and intel-lectual ability. It rein-forces the awareness of its bearer's self-worth, and enhances his sexual awareness and sexual attractiveness to other people.

It helps a person over-come past traumas that were difficult to cope with. It is an excellent stone for protecting the respiratory system and relieving problems arising from it. It is good for asthma sufferers, and helps cleanse and eliminate things that interfere with the functioning of the respiratory system, such as grains of dust, smoke, etc.

The stone also helps in treatment of digestive dis-turbances, nausea, and the liver and kidneys.

Crystals and Stones

Rhodonite

This stone balances the mind. It is important for reinforcing self-confidence and increasing self-esteem. It helps the bearer reveal the self, his inner ability, and his hidden potential, and helps him deal with long-latent emotions. The bearer has a good feeling and will hold his head up proudly as if in triumph.

This stone is soothing, and is good when dealing with problems concerning romance, or love in general. It repels negativity and alleviates feelings of stress and anxiety.

The stone intensifies the senses, especially the auditory system. It is particularly good for people who are in the field of music. Because of its general beneficial effect on the ears, it can be used for healing earache or ear infections.

Rock crystal

This is a transparent stone primarily used for foretelling the future. Crystal balls are made from it. This stone eases childbirth and strengthens the relationship between partners. It brings about situations of understanding and empathy with others. The stone brings luck as well as spiritual and mental calm to those using it. It is often carried in the pocket during weddings.

Crystals and Stones

Ruby

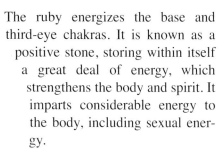

The ruby energizes the base and third-eye chakras. It is known as a positive stone, storing within itself a great deal of energy, which strengthens the body and spirit. It imparts considerable energy to the body, including sexual energy.

Because this is a powerful stone, it should be used cautiously.

If you feel inner power when holding a ruby, this stone is good for you. But if you are irritable and given to outbursts of rage, do not wear this stone.

The ruby is also used to develop a network of telepathic communication as well as communication with spirits. It has been attributed the properties of a safeguard against epidemics and evil spirits.

The ruby enhances self-awareness and connects between the Higher Self on the one hand and the subconscious and the conscious on the other.

The ruby symbolizes special qualities of giving and love between a couple and the ability to express emotions openly and without inhibitions.

The ruby is usually worn on a long chain so that it lies over the heart, or on a belt in order to strengthen the base chakra. In this spot it also affects the intestines, stomach, and of course the sexual and reproductive organs.

This stone is very useful in galvanizing a person

Crystals and Stones

into action because of its being saturated with energy, which is released into the body of the bearer. People who are prone to over-excitement, rashness or impulsiveness should therefore not use it.

This stone is helpful in releasing stress and in smoothing over quarrels. Anyone carrying it should know all its properties and be aware of its nature in order to elicit the maximum benefit from it.

The ruby crystal is also known as the "stone of honor," bestowing honor upon its owner, gathering and enhancing energy, and ensuring happiness and material success.

This is a very powerful stone and it has a great deal of energy. It has always been considered a stone with properties of strength, power, passion and fierce desire, with the help of which any obstacle can be overcome. The stone creates in its bearer a feeling of great inner power. If you do not feel good when holding it, and you feel that you are tending toward irritability and angry outbursts, do not wear it.

The stone is considered to clear and purify negative thoughts, and connects the bearer with positive feelings of love and limitless giving. Along with this, the stone banishes feelings of frustration and exploitation. The stone neutralizes emotions of unhappiness and disappointment and imparts a feeling of light and contentment.

This stone strengthens and balances everything connected with the heart: It dictated the pulses of a healthy heart and helps heal many cardiac diseases or ailments that derive from cardiac malfunction. It is best to wear the stone over the heart in order to allow it to radiate its positive force into it.

Crystals and Stones

It is good for balance between the body and the spirit: On the one hand, it activates the intuition and clairvoyant vision, while on the other hand, it helps to actualize ideas and practical objectives, and to implement them.

The ruby is a very spiritual stone and can help in many activities, as long as they stem from pure intentions and are free of the desire for self-aggrandizement.

It helps bolster a low self-image and develops leadership ability, so long as this is being done with spiritual inspiration and is not a result of egoistic intentions. The ruby helps create a feeling of self-fulfillment and aids in the absorption of positive energies.

When placed on the third eye during meditation, this stone helps a person evoke events or scenes from past lives.

Sapphire

The sapphire comes in various colors, ranging from yellow to black. It creates a feeling of great serenity in its bearer. It enhances his communicative abilities, improves his mood, reduces stress, and dissipates tensions and irritability.

It stops bleeding, improves vision, heals wounds, and allows a person to reach the true recognition of his nature and that of the universe.

The sapphire changes in color when it senses impending danger, thereby giving a warning. The stone helps a person control his urges, be they sexual or material. When placed on the third eye, the stone allows the conscious to penetrate to the depths of the subconscious.

Crystals and Stones

The sapphire enhances the inner wisdom of every person and suppresses negative emotions such as jealousy and envy. The stone augments inspiration and allows it to penetrate deeply into the intellect.

It allows one to see beyond blockages and is therefore excellent for healers. It permits the removal of blockages that hold back energies and interfere with their natural flow.

The sapphire is good for strengthening vision and reducing fever.

Blue sapphire: Helps in getting out of difficult mental states such as depression or deep despair. It is known for its ability to change a person's luck. It exerts a strong spiritual influence, enhancing its bearer's intuition and helping him become more attuned to himself and his environment.

Star sapphire: This stone is able to transform the energy systems in a person's body for the better. It is capable of creating balance, wisdom, and stability in the bearer.

Sard

This stone is good for concentration, for raising self-awareness and for intention. It is beneficial for treating hemorrhaging, protects against all sorts of bites and stings, and prevents ulcers and exposure to infections.

Selenite

As selenite consists of 70% water, it is relatively soft, and is therefore highly absorbent, absorbing both positive and negative energies. This is why it is usually used when dealing with emotions. It shifts energies, and just as it is able to absorb things, it is also able to transmit and radiate them back to the person.

This stone is good for improving memory, and it

Crystals and Stones

works in a positive manner on the female systems. It calms irritability, restlessness, and hyperactivity.

Selenite strengthens the immune system and the body's fluids, including the brain cells. When it appears in the form of rods, it is effective in the development of telepathic abilities.

Selenite reduces problems of calcium loss, alleviates the pain of joint inflammation, and strengthens bone structure.

Because this stone is very absorbent, it is usually programmed in such a way as to transmit information - thoughts and speech.

Some people whisper the material needed for a test into the stone and then put it under their pillow or into their pocket; when they need the information, they close their eyes and inhale through their noses.

By doing this, they allow the stone to lead them to the required information via a mental photograph of the page or by photographic memory.

Selenite reduces the damage caused by the effect of permanent metals in the mouth (bridges, fillings, crowns), promotes rejuvenation of the cell structure, and prevents the entry of free radicals into the cells.

Selenite is said to be a good stone for treating various forms of cancer and for reducing the sensitivity to light (for albinos as well).

This stone enhances the flexibility of the body's tendons and ligaments, helps balance, and prevents distortions in the structure of the skull; it also helps in cases of imbalance due to epilepsy.

Selenite aids in decision-making, in correct judgment, and in the communication between the conscious and the practical, occult and universal sides of the conscious and subconscious, thus permitting understanding of the true profound significance of any situation.

This stone is usually rounded in form. Sometimes it appears as a "V", and is then known as "fish tail."

Serpentine

This is a gentle stone whose greatness lies in its physical healing capabilities. It is a soft stone that imbues its bearer with a good, pleasant feeling. This stone alleviates states of tension and anxiety.

It provides protection against unpleasant surprises and helps get through periods of crisis in which the person feels that he is worthless and a loser.

It restores self-confidence and eliminates the feeling that the person is a victim of circumstances.

This stone balances the hormones and is recommended for post-partum women or nursing mothers, as it balances and increases milk production.

Some claim that the stone can protect people from snakebites and insect stings. It acts on the heart and on the lymphatic and respiratory systems.

Crystals and Stones

Smithsonite

This stone is known for honing the senses. It has the ability to enhance and intensify the power of an experience. It tends to heal those in pain, particularly mental pain. It can help one get through difficult traumas or crises in life more easily.

The stone radiates softness and tenderness, and it is therefore recommended during a Caesarian operation or while giving birth. The calmness will be transmitted to the newborn at the beginning of his life on earth.

Sodalite

Because of its blue color, this stone is good for all the functions that are affected by the color blue: It is excellent for flow and communication. It is particularly good for overcoming all sorts of fears and anxieties. It is excellent for cases of mental tension and irritability, particularly if they are associated with pangs of conscience or guilt feelings.

Sodalite strengthens the body physically. It is so powerful that it has become known as a stone that helps cure malignant disease.

Crystals and Stones

It encourages creativity and the exploitation of abilities that have not come to the fore previously. The stone eliminates emotional blockages and permits better spiritual understanding. It is good for harmony and a feeling of oneness with the environment, the universe, and the people close to the bearer. It promotes a feeling of courage. It is a good stone for people who have trouble establishing verbal or physical communication.

This stone helps the person free himself of habitual negative thoughts, in that it creates openness in him to the extent that he is able to develop an objective and less emotional approach. It permits rational thought. The stone can fortify physical tolerance and sensuality. It is considered beneficial in healing from toxins. It energizes the circulatory system and enhances the body's metabolism. It balances the thyroid gland and strengthens the lymph ducts.

Crystals and Stones

Sugilite (Royal Lavulite)

This stone affects clarity of thought and the balance between the cerebral hemispheres. The result is calmness and serenity. It is good for disturbances caused by malfunctions in the brain such as epilepsy or states of mental imbalance. It affords protection against negative energies and in particular calms children. It helps neutralize fears and instills confidence. Sugilite can help children who have trouble with their schoolwork or experience difficulties with writing or reading, for reasons such as dyslexia, to improve their scholastic functioning.

It is beneficial in developing the sixth sense and affects the third eye and communicative abilities.

Tektite

This stone is a meteorite, a natural glass, sometimes called "star glass." It connects between spirit and body, and is considered to have a calming effect. It causes a person to be more grounded and logical, and less "spaced-out" and emotional.

This stone is considered to be a powerful amulet and is one of the stones recommended for communication with extra-terrestrial beings.

Tiger's-Eye

Tiger's-eye appears in various colors, each with its own unique qualities. In general, it can be said that it is good for anyone who needs to express himself in a variety of ways, such as actors, for instance. It enables its bearer to see reality in a different light or from different angles.

Red tiger's-eye: This stone enhances trust in the ego and the human self, enables the person to do things that he did not think himself capable of doing, and reinforces his faith in his own capabilities. It brings him to self-realization.

Blue tiger's-eye: This stone is capable of divine transmission. When placed on the third eye, the world and the things that happen to people can be surveyed from the proper perspective. It helps to put things in perspective. With its help, the person can look at life objectively.

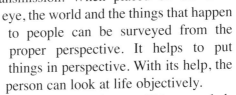

Golden tiger's-eye: This stone balances mental and physical needs. It helps eliminate guilt feelings and unwanted thoughts. It has a beneficial effect on the gall bladder, kidneys, and liver.

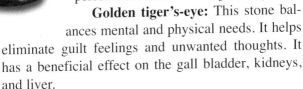

Topaz

This stone, due to its dominant yellow-gold-orange coloring, sends out warm, healing, life-giving rays, like the sun. It cures insanity and various types of mental illness, curbs hemorrhaging, heals infections, soothes and induces deep, undisturbed sleep, and serves as a provider of emotional balance and a fear suppressant. It is beneficial in cases of chronic fatigue or exhaustion. This is a stone with qualities of vitality and renewal. It signifies the rebirth of the genuine spiritual self. It increases physical energy and vitality, and is good for anything connecting with problems of eating and digestion: It improves the appetite and the sense of taste, and, alternatively, it is also good for those who wish to lose excess weight. It helps heal aging-related ailments or diseases which necessitate tissue rejuvenation and the healing of wounds.

Topaz eliminates lack of clarity and gives its bearer the opportunity to look at his life clearly and not hazily. It is good for stabilizing problems concerning the family. It is excellent for those whose principal occupation is solving logical problems, as it will help them clarify things and reach the solution more easily. Topaz is good for those who need to improve their speaking or communicative abilities. It is excellent for speechmaking, and

affords protection against dangers, black magic, and accidents.

Blue topaz: This stone enhances its bearer's self-control and leadership qualities if he possesses any. It exerts a good influence on people who insulate themselves against receiving help from those around them or against things given them by people close to them, such as gifts from members of the family or the like. The stone releases tensions and inhibitions, and is known to help men who wish to control their ejaculation during sexual intercourse. This stone is linked to the voices of the universe.

Tourmaline

This stone affects almost all the chakras, meaning all the energy centers in the body, and it is therefore good for general use, physical and spiritual alike. Tourmaline has electrical and magnetic properties when rubbed or heated. It enhances telepathic ability and awareness, and helps to focus thought and concentration.

Tourmaline appears in a wide range of colors, each with its own qualities. In general, it can be said that pale tourmalines are good for focusing spiritual energy in healing, relaxation, and meditation, while the dark stones have more of a physical effect.

Tourmaline is an "obligatory stone" for anyone practicing healing and for everyone who believes in stones and uses them.

Crystals and stones

Black tourmaline: This stone protects the person from negative influences and the evil eye. It is a grounding stone and enables its bearer to be realistic and to integrate his spiritual self into his physical body. Because of its quality of not absorbing negative energy and influences, it is often carried as an amulet which can only help.

It is effective against debilitating diseases or chronic diseases such as heart disease or rheumatism. It also strengthens the immune system.

Black tourmaline intensifies whatever mood its bearer is in, so it should not be carried when the person is in a bad mood or is feeling dejected or irritable.

Watermelon tourmaline: The combination of the green color around the red is reminiscent of a watermelon. It is known that reddish/pink and green are complementary colors.

This stone helps release old fears. It brings harmony between opposites and creates balance between yin and yang. It is good for use between partners.

Pink tourmaline (rubellite): This stone strengthens creativity and sexuality. It is good for internal balance and for strengthening the ego and self-confidence. It stimulates love and strengthens will power.

Blue tourmaline: This stone helps improve communication between its bearer and those around him. It is very good for curing depression and easing mental stress; it calms,

and helps a person relax from tension and flow with life. It frees him of guilt feelings and helps him be less critical and more forgiving toward himself.

Green tourmaline: This stone exerts a very strong influence on health. It is considered a healing stone with the ability to affect every system in the body. It balances the energies in the body, calms nerves, and dispels fears. It works positively on the brain and strengthens the immune system.

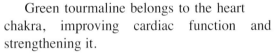

Green tourmaline belongs to the heart chakra, improving cardiac function and strengthening it.

It is excellent for people who want to turn over a new leaf in life or who are facing new beginnings and innovations. It helps people get out of difficult spiritual and mental situations and cope with problems and new challenges.

White tourmaline: This stone has an effect on two opposing planes: It affects the digestive system, the spleen, the white blood cells, and more. On the other hand, it affects the person's spiritual and emotional systems.

Cat's eye tourmaline: This stone strengthens the sixth chakra - the third eye. It develops clairvoyance, as well as the power to see the occult and foretell the future. It is used for contact with higher beings and develops the ability to see ahead. It is good for use in meditation.

Quartz tourmaline: This stone has many properties. It is capable of charging the body with a great deal of energy; thus it is useful in situations of feebleness and weakness. It dispels fears and anxieties and exerts a beneficial influence on the nervous system. It is good in cases of emergency.

Turquoise

This stone is considered to bring great luck to its bearer. It is known as a stone that wards off disasters and calamities. Considered a "personal" stone, it creates a unique relationship between the giver and the recipient when it is received as a gift or given to someone else as a gift. It brings a special blessing to the recipient. This stone provides protection and imparts serenity and calm.

Because of its blue and green coloring, turquoise serves as a very powerful healing stone with considerable influence. It is effective in healing throat problems, lung and heart diseases, eye infections, and various other infections.

This stone keeps away demons and night spirits and neutralizes spells. The custom of painting the doorways to houses blue originated from the turquoise.

This stone helps in the absorption of nutrients in the blood and in problems of malnutrition. It influences all the energy systems of the body, strengthening and balancing them.